The
Model Engineer
Series
No. 15

Net

Simple
Scientific Experiments

*How to Perform Entertaining and Instructive Experiments
with Simple Home-made Apparatus*

BY

AUREL DE RATTI

FULLY ILLUSTRATED

SECOND EDITION

Describes Forty-six Instructive Experiments in Electricity, Magnetism,
Hydraulics, Hydrostatics, Light, and Acoustics.

Simple Scientific Experiments

Originally published by
Percival Marshall & Co.
London

Reprinted by
Lindsay Publications Inc
Bradley IL 60915

ISBN 1-55918-266-0

2 3 4 5 6 7 8 9 0

2002

The "Model Engineer" Series.   No. 15

SIMPLE SCIENTIFIC EXPERIMENTS

ENTERTAINING AND INSTRUCTIVE

A Practical Handbook of Effective Experiments in Electricity, Magnetism, Hydraulics, Hydrostatics, Light, Acoustics, etc., with Simple Home-made Apparatus

BY

AUREL DE RATTI

FULLY ILLUSTRATED

SECOND EDITION

LONDON
PERCIVAL MARSHALL & CO.
26-29 POPPIN'S COURT, FLEET STREET, E.C.

CONTENTS.

PREFACE.

THE object of this book is to provide scientific amusement by means of simple and inexpensive apparatus. As most of the experiments here introduced are, if not quite original, at least original in the form which I have given them, the book may perhaps also be found a useful adjunct to the ordinary Science text-book. Having often performed the experiments in my own lectures and with apparatus made by myself, I feel confident that the directions I have given for making the apparatus will enable any one possessed of a little mechanical skill to construct the necessary apparatus for himself, and that the experiments themselves will be found a source of real pleasure, and not, as is often the case, "an idealisation on paper and a disappointment in practice." I assume, of course, that the student is endowed with sufficient patience, for want of which I have often known experiments to fail in the hands of people who in their eagerness to obtain the desired effect neglected to carry out the directions laid down for their guidance.

A very small sum will supply all accessories required (with the exception of an induction coil), and in cases where access to a laboratory can be had, practically no expense will be incurred.

AUREL DE RATTI.

HARROGATE.

SIMPLE SCIENTIFIC EXPERIMENTS.

EXPERIMENT NO. 1.

Model Hydraulic Ram (fig. 1).—This apparatus, which was invented by Montgolfier, is very instructive. A model can easily be constructed with

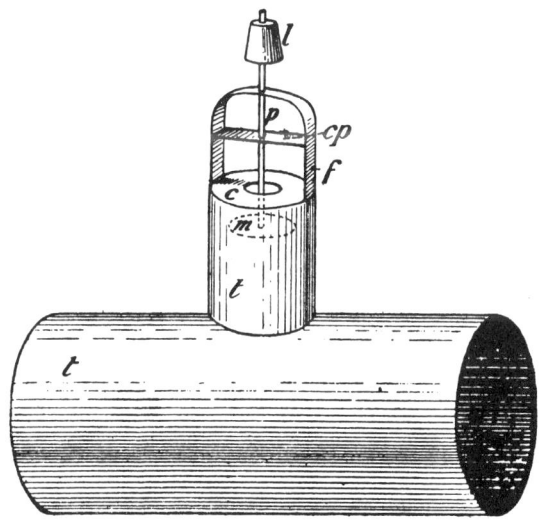

FIG. 1.

the help of a few inches of metal tubing made of an old meat tin. (1) Join two pieces of tubing, $t\,t$, as shown in the figure, $\frac{1}{3}$ to $\frac{1}{2}$ of an inch in diameter. Close the short tube with a cover, c, which has a hole cut into it so as to allow the piston and rod, p, to work freely up and down. The piston itself is made of a circular piece of metal, m, about two-thirds of the diameter of the tube, and must be perfectly level, as must also be the underside of the cover, so that when the two surfaces come into contact the hole should be completely closed to the flow of water. In order to regulate the movement of the piston a small frame, f, with a cross piece, cp, is soldered to the

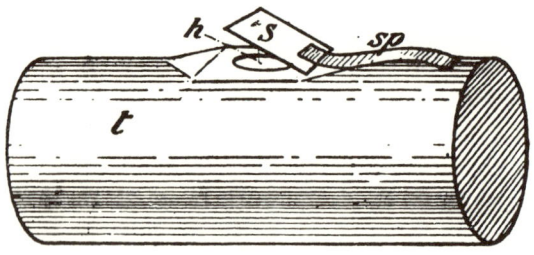

FIG. 2.

tube. The piston rod is passed through the holes in the frame, and a small piece of lead, l, or other metal, doubled over to the end of the rod will hold the latter in position and prevent the piston from descending too far; it will also act as a regulator, and must therefore be movable, but firmly remain in any position. (2) Fig. 2 : Make a small valve of a piece of tubing, t, $\frac{1}{2}$ in. by $1\frac{1}{2}$; this must be closed at one end and have a hole, h, of about $\frac{1}{4}$ inch in the middle. Over this hole is placed a small piece of sheet lead, s, which is held in position by a weak spring, sp, made of a copper strip (or a piece of a watch spring will answer), one end of the spring being soldered to the end of the pipe, the other to the piece of lead. The spring must just be strong enough to keep the lead on the hole and be elastic enough to allow of a movement of 45 degs., and the valve must fit accurately on the hole so as to prevent the flow of water when closed. (3) Fig. 3: Connect the two pieces of apparatus by

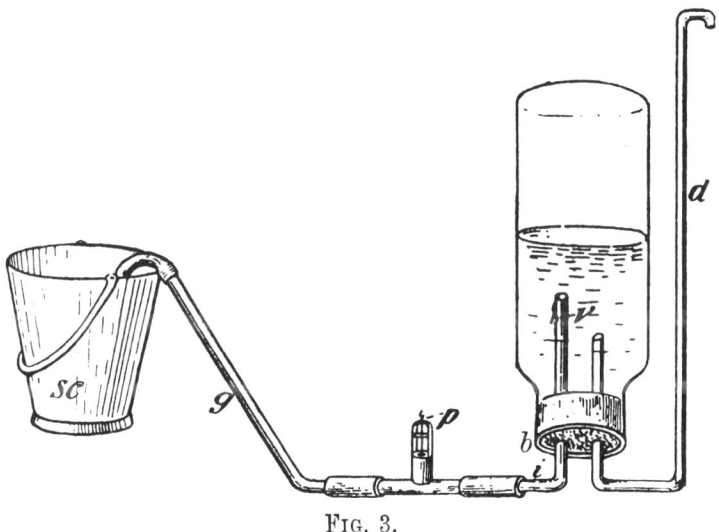

Fig. 3.

means of a few inches of *thick* indiarubber tubing, *i*, such as will not expand under pressure. Of course metal tubing—ordinary gas-pipe—soldered to the two pieces will answer as well. Previous to join- ing the two pieces the valve, *v*, is passed through a cork, *b*, fitting tight into the mouth of a bottle; the cork has a hole for the insertion of a delivery tube, *d*. This tube may be short and bent up, ending in a point if a fountain is to be produced;

or rise several feet above the level of the supply cistern, sc, to demonstrate the raising of water from a low to a higher level. (4) Join the tube carrying the piston by means of indiarubber tubing or gas piping, g, to a reservoir about 2 ft. above the piston. (5) To set the apparatus to work, pull up the piston, p, thus shutting up the piston outlet hole, and allow the water to flow into the bottle. Now close the delivery pipe, d, and work the piston up and down alternately, allowing some water to flow out, and then abruptly shutting the hole. Every time this is done the moving column of water will open the valve, v, in the bottle, a quantity of water will be forced into it, and the air above it will be compressed. With proper adjustment the apparatus will soon work of its own accord. Only very little play is required for the proper action of the piston—the necessary adjustment must be found by practice, but the apparatus will not work unless the valve in the bottle opens wide *so as to allow a back flow of some water* after each stroke. This back flow exerts a sucking action upon the piston, p, and causes it to fall, which it would not do otherwise as the pressure of the water would keep it shut. Most text-books state that the piston falls by its own weight, but it must be obvious that it cannot do so.

Experiment No. 2.

A Simple Water Turbine (fig. 4).—In most cases where the power of waterfalls is utilised for obtaining energy to be transmitted in the form of electricity, turbines are used in preference to ordinary water wheels.   A model may easily be constructed

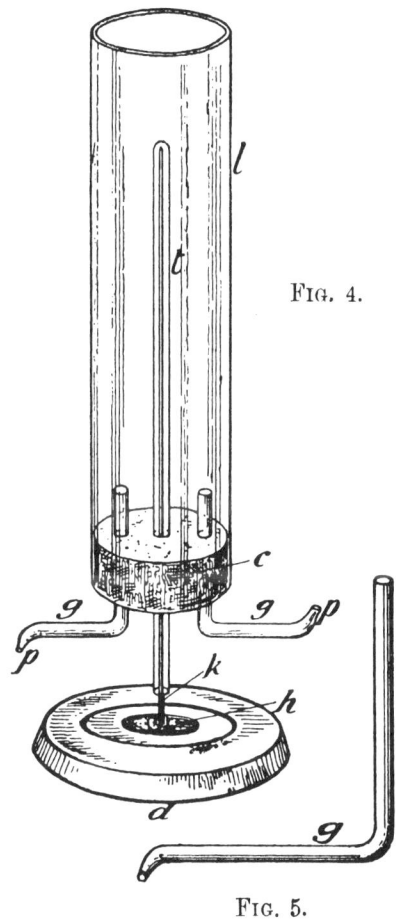

Fig. 4.

Fig. 5.

of a glass lamp chimney, *l*, 8 or 10 ins. long by about 2 wide, fitted with a cork, *c*, at one end.   Two glass tubes, *p p*, twice bent at right angles and drawn out at the ends, are passed through the cork, *c*, as shown in the drawing (fig. 5). Another glass tube, t, closed at one end, is passed through the middle of the cork with the closed end inside the chimney. A strong knitting needle, k, is fixed upright on a board or on a wooden disc, d, such as are used by plumbers for fixing gas brackets, by means of a cork inserted in the hole, h, of the disc, and the centre tube is slipped over the needle. All parts must be carefully fitted, or the chimney will not turn true. To work the apparatus pour water into the chimney, l, when the reaction of the water flowing out at the points, p, will cause the apparatus to rotate. Special care must be taken in fitting the cork, c, as it is apt to break the glass ; it is best to select an easy fitting cork and cement it to the glass with plaster of Paris or cycle cement. Instead of the glass chimney a narrow canister may be used with a hole bored through the bottom for the centre tube, which may be cemented to, or passed through, a small cork fixed in the hole. The two delivery tubes are preferably passed through the sides, and of course need only be bent once at right angles.

The apparatus may be simplified by suspending the tin canister from a string (fig. 6), fastened to a small swivel, s,

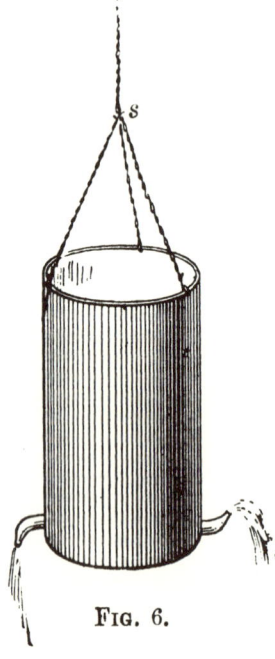

FIG. 6.

such as are used for fishing lines. Such an arrangement is very effectual, and illustrates the reaction of
water flowing from an opening per-
fectly well. The tin may also be
placed on a board floating on the
surface of water; in this case the
board will turn with the tin, but
the rotation is slow and the experiment less effective.

EXPERIMENT No. 3.

The Syphon (fig. 7) illustrating
the principle of the apparatus. Pass
a short, s, and a long, l, piece of
glass tubing through a cork, c, which
is tightly fixed into the neck of a
bottle, p. The end of the short
tube must be drawn out to a point,
p, with a small opening. Insert
the short tube in a vessel of water,
w, and suck the air from the long
tube, l. The water will rise through
the small tube in the form of a
fountain and flow away through the
long tube. The upper end of the
long tube must of course be below
the point, p, of the short tube. If
now, when the apparatus is working,

FIG. 7.

the flow of water be arrested by putting a finger on
the outlet, o, the fountain will continue to play for a

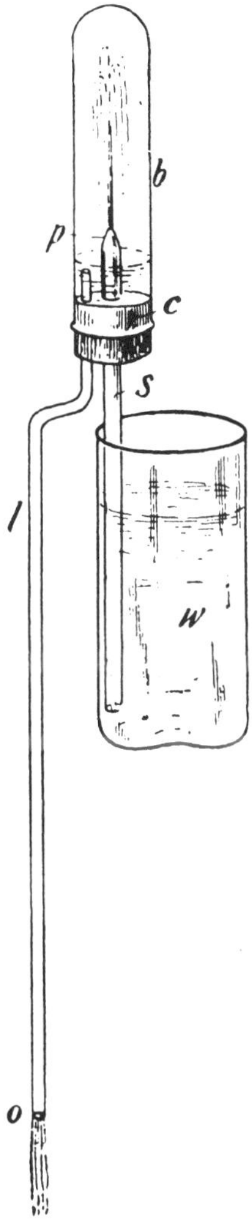

few seconds, showing that a partial vacuum is produced in the bottle, and proving the theory of the
syphon.

EXPERIMENT No. 4.

A Pretty Experiment with the Syphon (fig. 8).
—Draw out a piece of glass tubing, *t*, 4 ft. long, to

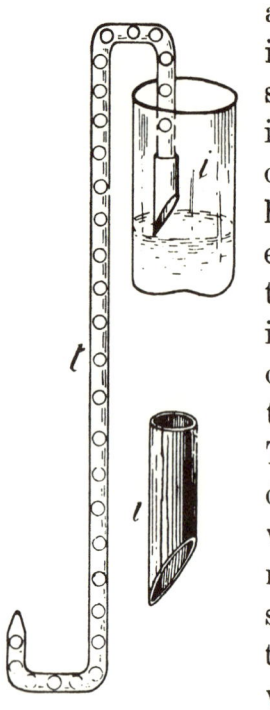

FIG. 8.

a point, bend it twice as shown
in the figure, carefully avoiding
sharp angles.   Slip an inch of
indiarubber tubing, *i*, over the
other end, and cut the projecting
half off obliquely.   Place this
end in a vessel of water, start
the syphon, and now support it
in such a way as to raise the
oval opening of the indiarubber
tube a little out of the water.
This will cause the water to
draw in bubbles of air, which
will pass down the tube alternately with drops of water.   By
slightly raising or lowering the
tube the character of these spaces
will change.   The water should
be coloured with a little aniline.

If the syphon is lifted out of the water and 10 or 12
inches of air are allowed to enter before it is lowered
again, this long bubble will pass on slowly until it
arrives at the outlet, when the air will pass out with
great velocity.   The liquid behind acquires the

same velocity, and on reaching the point it will be ejected with considerable force, rising for a moment to a height of 6 or 7 ft. The same action takes place in the hydraulic ram.

EXPERIMENT No. 5.

Electric Fountain (fig. 9).—A simple, easily arranged and striking experiment. Fix a small

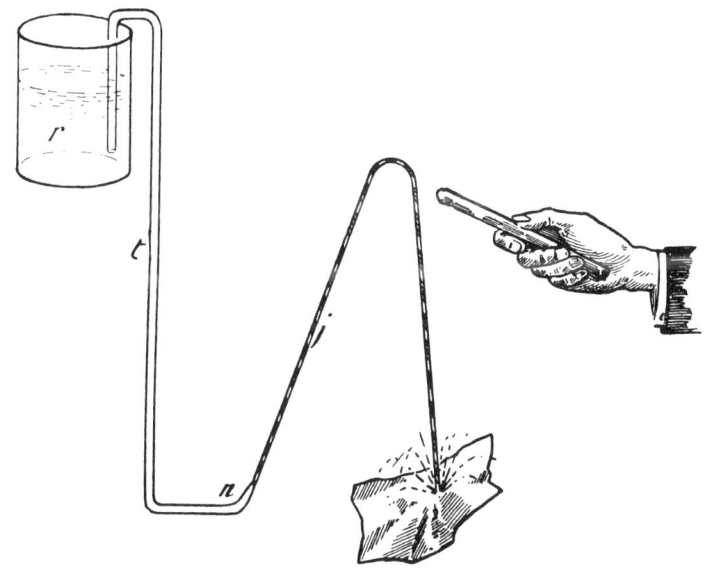

FIG. 9.

nozzle, n (made of a piece of glass tubing drawn out), in nearly upright position and connect it by means of indiarubber tubing, t, to a reservoir, r, about 3 ft.

above the nozzle. Cause the fountain to play; the issuing jet will be slightly inclined to one side. Now rub a piece of sealing wax, vulcanite, or a stick of sulphur on your coat sleeve and hold it opposite the jet of water at a distance of a few feet, when the scattering drops of water will at once unite and fall in a solid stream. On removing the sealing wax the jet will assume its ordinary appearance. If the water be allowed to fall on a piece of stiff paper, a difference in the sound, caused by the water either falling in a stream or in drops, will be noticed. To perform this experiment successfully, care must be taken to have perfectly dry hands and dry sealing wax.

Experiment No. 6.

Acoustic Fountain.—Use the same arrangement. Rest one end of a thin walking stick, yard measure, or similar length of wood firmly against the nozzle. Vibrate a tuning fork and apply its lower end to the other end of the stick. The jet of the fountain will divide into two, three and sometimes more distinct jets as long as the fork is sounding.

Experiment No. 7.

Glass Balls on Fountain Jets (fig. 10).—To perform this experiment successfully the jet must be perfectly upright. Place a glass bead of about $\frac{1}{2}$ in. in diameter, or a table tennis ball b, on the jet. The bead or ball will be carried to the top of the jet and there remain rotating around its own axis. Beads such as are used for decorating Christmas trees will answer very well, but the holes in them must be closed with a little wax, and care must be taken that they are absolutely free from grease or they will not remain on the jet but fall off. To free a bead from all grease leave it for a few minutes in spirit of wine, or wash it with strong soap. A piece of funnel-shaped wire netting placed above the nozzle will catch the ball if it happens to fall off; the ball will roll to the centre and be caught again by the jet of water.

Fig. 10.

EXPERIMENT No. 8.

Vortex Rings (fig. 11).—An oblong biscuit tin, b, 10 by 4 or 5 ins., will be found most suitable for this experiment. Cut a neat round hole, h, about 2 ins. diameter, in one end of the box; remove the other end altogether. Replace the lid, and make it airtight by glueing a strip of paper, p, all round.

2

Now fix, by means of cycle cement, a sheet of indiarubber over the open end of the box; a piece cut from an old football bladder will answer very well, and it should be slightly stretched so as to rebound readily when struck. Fold up two pieces of blotting paper so that they can be passed through the hole into the box. Soak one piece with strong hydrochloric acid and the other with strong ammonia. Drop them into the box, pushing the first to the extreme end and keeping the other in

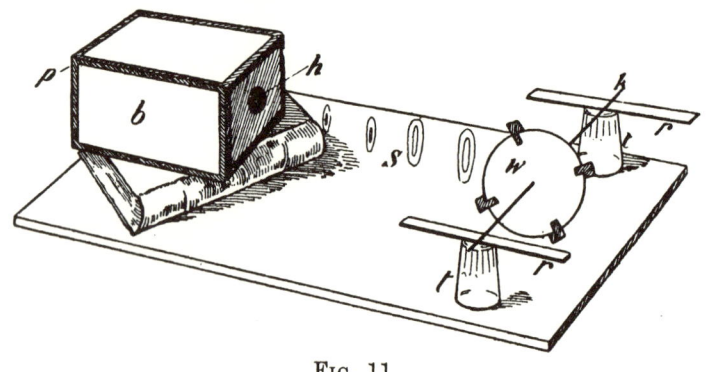

Fig. 11.

or near the middle of the box. They must not touch each other. A dense white smoke will at once fill the box, and by tapping the indiarubber an endless number of smoke rings, s, can be projected into space. With a little practice and in a quiet room it will be possible to show that when two rings follow each other with their planes parallel and their centres moving in the same line, the foremost will relax its speed and spread out, while the follower will quicken its pace till it passes through the other.   When one vortex ring impinges obliquely on another, it rebounds from it and both are thrown into vibration.   They act like solid indiarubber rings.   A vortex ring made to impinge on a candle flame will, even at a distance of some yards, extinguish it.   This experiment will also succeed without any smoke being used ; the effect in this case is very surprising.   A piece of paper loosely suspended from a string will be violently agitated if struck by a vortex ring.   A

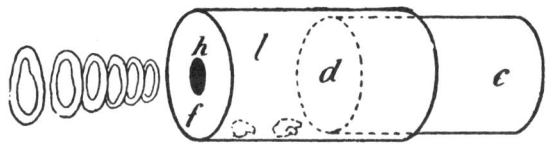

FIG. 12.

very striking effect is also produced by placing a small paper wheel, *w*, at a distance from the box ; under the impact of invisible vortex rings the wheel will rotate rapidly.   A paper wheel as shown in the figure can easily be put together ; a thin, perfectly straight knitting needle, *k*, forms the axle, the ends of which are placed on flat rulers, *r r*, supported by a couple of tumblers, *t t*.   A little sealing wax is used as cement to fix the wheel to the knitting needle.

EXPERIMENT NO. 9.

Modified Experiment with Vortex Rings.— The experiment may be modified in the following manner (fig. 12):—Stretch an indiarubber dia-

phragm, d,—a piece cut from a toy balloon will do—
over the end of a cardboard cylinder, c, about 4 ins.
in diameter, and secure it by means of thin wire or
string. Slip a second somewhat larger cylinder, l,
over the first one so as to increase the length of the
latter by a few inches at the diaphragm end. The
second cylinder is closed in front, f, with the exception of $\frac{1}{4}$ of an inch round hole, h. Pieces of
blotting paper charged with hydrochloric acid and
ammonia are placed between the diaphragm and the
aperture. If now sounds are uttered into the open
end of the cylinder, the vibration of the diaphragm
causes puffs of air to issue from the small hole, made
visible by the thick smoke of ammonium chloride
in the form of vortex rings. The sounds uttered
into the cylinder should be of a very low pitch, as
else the rings will merge into one another. The
small paper wheel can be made to turn round if
placed at a short distance from the opening. *See*
Experiment 8.

Experiment No. 10.

Explosiveness of Flour (fig. 13).—Make a small
hole near the bottom of a large biscuit tin or
similar box, pass the end of a length of indiarubber
tubing through it, and put a handful of dry flour in
front of the tube. Now place a short piece of
lighted candle in the box, put the lid on, and blow
through the tube by means of a pair of bellows or
with the mouth. An explosion will take place, and

the lid, which should not be very firmly fixed, will be blown off. Instead of flour, finely powdered resin, lycopodium, and even fine dust, such as collected

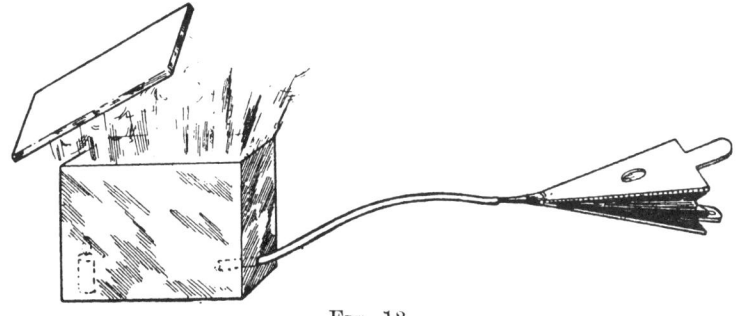

FIG. 13.

from elevated parts of rooms, on the back of pictures, etc., may be used. If flour is used, care must be taken to leave it for some time near a fire so as to remove every trace of dampness.

EXPERIMENT No. 11.

Spontaneous Combustion of finely divided Iron and Lead (fig. 14).—Close an ordinary piece of glass tubing 4 to 5 ins. in length at one end by heating in a Bunsen flame, fill it one third with either oxalate of iron or tartrate of lead. Now draw out the other end to a fine point, and when cold break the tube at the neck, thus leaving a small opening which should be not less than $\frac{1}{8}$ of an inch. The tube must now be held horizontally and its contents spread out evenly over the lower half by tapping it or shaking it. The heat from a Bunsen burner is now applied at the closed end and gradually over the whole tube. When the contents have turned black and vapour ceased to issue from the point, the latter is quickly closed by holding it in the upper part of the flame. The tube is laid aside to cool, and is then ready for showing the spontaneous combustion of its present contents—*i.e.* finely divided iron or lead, obtained by the decomposition of the salts of these metals. Scratch the tube in

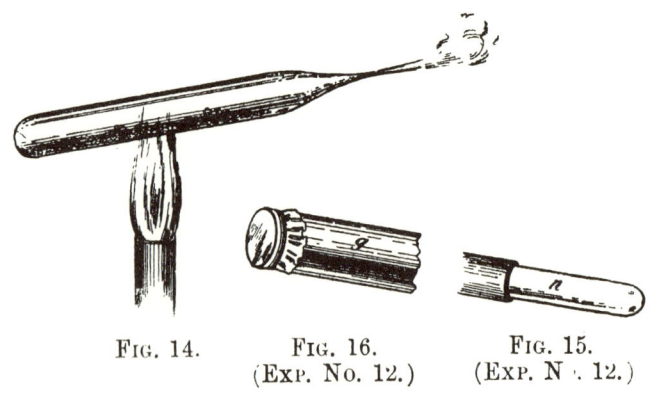

| FIG. 14. | FIG. 16. | FIG. 15. |
| | (EXP. NO. 12.) | (EXP. N . 12.) |

the middle with a file and break it, at the same time scattering its contents. It is best to do this from a position as high above the ground as possible. There will be a shower of sparks produced by the burning metal. Before breaking a tube it should be well shaken to loosen the contents.

EXPERIMENT No. 12.

Acoustic Water Jet (fig. 15).—This is a most attractive and striking experiment. Its success depends upon a properly shaped nozzle. The nozzle,

n, is made by heating the end of a piece of quill glass tubing in the upper part of a Bunsen burner, turning it constantly so as to heat it equally. After a few minutes the opening will begin to contract, and if allowed, will do so, until it is quite closed. This must, however, be prevented, as a small hole is required ($\frac{1}{50}$ to $\frac{1}{70}$ of an inch in diameter), and the object is best attained by gently blowing through the tube when the issuing blast of air in the flame will give sufficient indication of the size of the hole. The student should make half a dozen of these nozzles

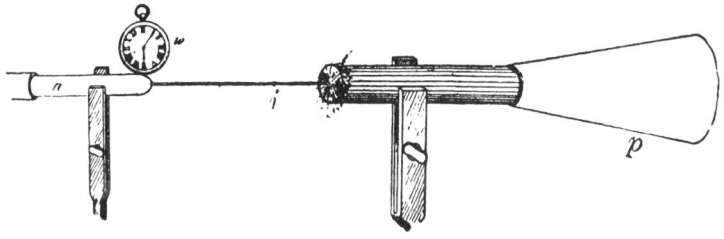

FIG. 17.

one or two will probably prove satisfactory. An ordinary nozzle, *i.e.* one drawn out to a neck and broken off, such as used in previous and similar experiments, will *not* answer. The next thing required is a small piece of thin sheet indiarubber such as is used for toy balloons. This should be stretched, but not too much, and tied over the end of a metal or glass tube of $\frac{1}{2}$ to $\frac{3}{4}$ of an inch in diameter (fig. 16, g). The other end of this tube is slipped into the small opening of a paper cone, p, fig. 17, which is firmly held there by a few turns of string. This cone may be some 8 or 10 ins. long.

The tube with the cone must now be fixed in a horizontal position, or it may be held with the hand during the performance of the experiment. A bucket of water some 10 or 15 ft. above the nozzle n is connected with the latter. It is best to firmly fix the nozzle in a clamp in a horizontal position so as to obtain a steady jet of water, j, which is to be directed against the sheet of indiarubber, i. At a short distance of, say, 6 or 8 ins. the jet will impinge on the elastic skin without producing any sound, but as the distance is gradually increased a shrill whistling sound is given out by the cone. For our experiment the distance should be such for the two parts as to cause the jet of water just to become audible. If the nozzle be now slightly tapped with a match or other light object the sound will be magnified to such a degree as to be audible all over a large room. The friction of a small file drawn loosely over the nozzle will likewise produce a surprising effect. The ticking of a watch, w, placed on, or held against, the nozzle, becomes very audible to a great distance and resembles the noise of hammering if the apparatus is properly adjusted. Care must be taken to have perfectly clean water free from small particles floating about in it. Direct connection with the house supply cannot be recommended on account of the constant change of pressure in the mains. An equal pressure is absolutely necessary. If the nozzle should by any means become choked it can easily be cleared of the obstruction by allowing water to flow through it in the opposite direction. The piece of indiarubber should be cut from a bladder that has not previously been blown up.

EXPERIMENT NO. 13.

Manufacture of Gas (fig. 18).—Experiment to show the principle of coal gas manufacture. Roll up a large sheet of paper into a cylinder, c, of about 2 ins. diameter. Flatten this cylinder by placing it on the table and pressing it down. Double up one end, e, so as to close it up securely, and cut two or three holes about $\frac{1}{8}$ of an inch diameter

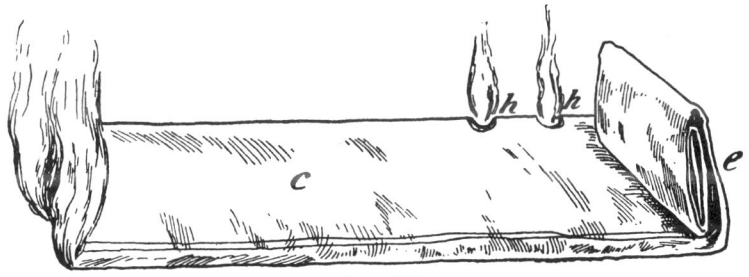

FIG. 18.

along the edge near the end, taking care that no layer of paper remains uncut. Now hold the roll straight before you, the holes being somewhat higher than the open end. Light the open end well so as to produce a large flame. A quantity of the inner layers of the roll will not burn but undergo destructive distillation, and the unconsumed gas will travel up the tube and issue in the form of thick white smoke from the small holes. This gas can be lighted, and will burn like a gas flame. The lower part of the issuing gas does not, as a rule, ignite, and the paper remains therefore intact to the end.   Some kinds of paper do not lend themselves to this experiment, especially those containing much mineral matter.   The *best* quality of brown paper answers well, as does the finest quality foolscap. A suitable paper must be found by trying.   Newspapers are useless.   Small pieces of glass tubing may be inserted in the holes h to act as burners.

EXPERIMENTS NOS. 14 AND 15.

Two Experiments with the Induction Coil* (fig. 19).—(1) *Fluorescent Writing.*   Among the many experiments with the coil and fluorescent liquids

FIG. 19.

the following one seems to be little known.   Trace a design or write a few words with sulphate of quinine on a white piece of paper.   When dry, the writing will be quite invisible, but on exposing

* For instructions on making an experimental induction coil see No. 11 of *The Model Engineer* Series of Handbooks, and for batteries see No. 5 of the same series.

it in a dark room to the light of an ordinary Geissler tube, or that produced in an electric egg, the writing or design will appear in blue colour.

(2) *Imitation of Gassiot's Cascade* (fig. 20).——Tie a wire, *w*, round a cylindrical bottle, *b*, about half way up, and connect it with one pole of the coil. The wire from the other pole, *w w*, is put inside the

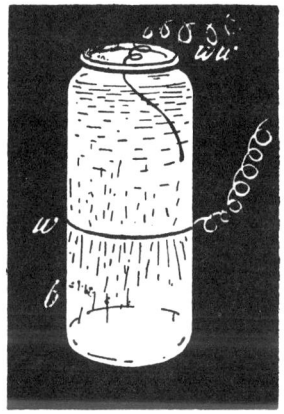

FIG. 20.

bottle and the end of it immersed in a small quantity of water in the bottle.   On the coil being set to work the whole outside of the bottle will be covered with a network of sparks, starting from the wire and darting in all directions.   This experiment resembles Gassiot's cascade, but requires no vacuum.

EXPERIMENT No. 16.

Effect produced by Persistence of Vision (fig. 21).——A slender glass rod is bent in a Bunsen flame so as to conform with the outline of one of the sides of the sketch of a vase or similar ornament, v. Along the whole length of the tube is cemented a very narrow strip of tinfoil, which is divided at short intervals with the point of a sharp pen-knife. By an arrangement as shown in the figure the tube is supported so as to revolve freely in an upright position, its lower end being cemented to the centre of the small metal pulley, p, of a whirling

FIG. 21.

table, t, while its upper end is held in position by a wire, w, projecting from a wooden arm, a. This wire passes a short distance into the glass tube v. One of the poles from an induction coil is connected with this wire, while the other is resting against the pulley. The glass tube is now set revolving, and the current passed through the tinfoil, illuminating all the cut spaces.

If the glass has been skilfully bent and the discharge is sufficiently rapid, the effect is one of startling beauty. A Wimshurst machine may be substituted for the coil.

Persistence of vision transforms the revolving luminous glass tube into a perfectly symmetrical figure, glimmering phantom-like out of darkness, and covered with thousands of flashes of shimmering light, having the appearance of the most delicately fashioned wirework, wrought in curves of sparkling fire.

A small motor driven by one or two bichromate cells may with advantage be used instead of a whirling table. The motor is fixed to a baseboard in such a position as to have its revolving shaft pointing upwards, and the lower end of the glass tube is slipped over the end of the shaft and secured with cement, the upper end being held in position as described. Where difficulties are met in obtaining a properly bent glass tube a straight tube with a spiral strip of tinfoil may be used, but the effect produced is of course inferior.

EXPERIMENT No. 17.

Aspirator and Blower (fig. 22).—Fit two corks, $c\,c$, to a wide glass tube, t, and pass a narrow glass tube through each, one of them being somewhat narrowed at the end (fig. 23), e, the other having a neck, n, and being somewhat widened so as to allow of the insertion of the other. The proper shape can be easily obtained over a Bunsen flame with the help of a pair of small pliers. Another piece of glass tubing, *r*, bent at right angles, admits air into the large

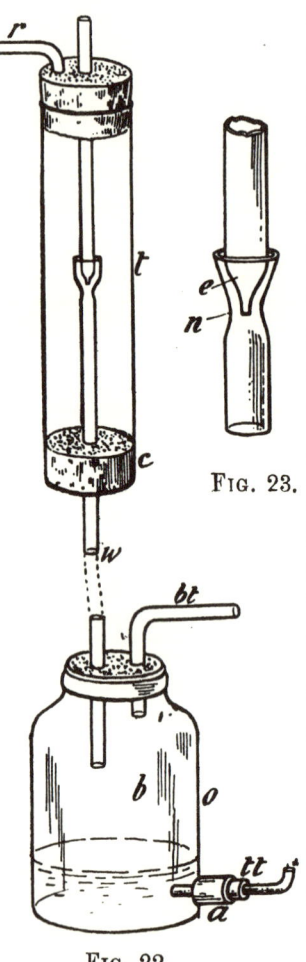

FIG. 23.

FIG. 22.

tube, *t*, and to it is attached a piece of indiarubber tubing sufficiently stout not to collapse under the atmospheric pressure. The distance between the point and funnel of the inner tubes must be found by practice. With a good pressure of water a surprising degree of exhaustion can be obtained. The air carried along by the issuing jet of water, *w*, can be collected in a condenser, *o*, and employed for producing a blast useful for blow-pipe purposes, drying, blowing soap-bubbles, etc. For constructing an efficient condenser, a bottle, *b*, with an opening or tubulure, *a*, at the bottom is necessary. A glance at the engraving will show this arrangement. The opening of the tube, *t t*, for carrying off the waste must be just large enough to allow the water to issue with some force, so that there may always be a quantity of water retained in the condenser under pressure. The air carried down from the aspirator issues through the bent tube, $b\,t$, at the top.   The force of the blast depends upon the vertical distance between the aspirator and condenser.

EXPERIMENT No. 18.

Static Electric Motor (fig. 24).—Slip a small cork, c, over a knitting needle, n, and insert in the cork at right angles four thin spokes, s, of glass or

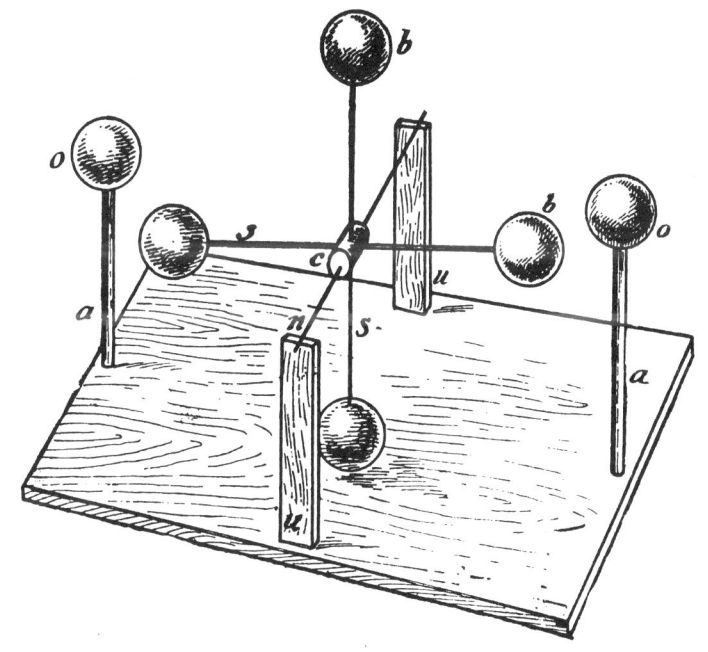

FIG. 24.

vulcanite carrying each at its end a light ball, b, an inch or more in diameter.   Most suitable for the purpose will be found small celluloid balls such as are obtainable in any toy shop.   The balls should be carefully covered with Dutch metal or very thin tin-foil. Now support the knitting needle between two uprights, u, so that the balls will rotate easily on the slightest impulse. At right angles to the uprights, u, fix two more balls, o, on insulated uprights, a. These balls should just allow the rotating balls to pass quite closely without touching them. Connect each of the outside balls o with one of the conductors of an electrical machine, when, on the machine being worked, attraction and repulsion will cause a rapid rotation of the spindle. If only one kind of electricity is available the arrangement must be so modified that the receding ball on passing the stationary ball gives up to it all electricity through the medium of a small piece of tinfoil projecting from the latter, and the ball must of course be connected with the ground. The attractive and repulsive force of static electricity is far from powerful, and the machinery operated by it should therefore be light, well balanced and easy running.

Experiment No. 19.

Curious Thermo-Magnetic Motor (fig. 25).—This motor, though devoid of practical value, is of sufficient scientific interest to warrant a description. A disc or ring, r, of thin iron or steel is mounted on an axis so as to move freely. The edges of the wheel are placed opposite the poles of a magnet, m. If now any part of the wheel between the poles be sufficiently heated it will slowly rotate round its axis.   The   heated   part   being   less   powerfully

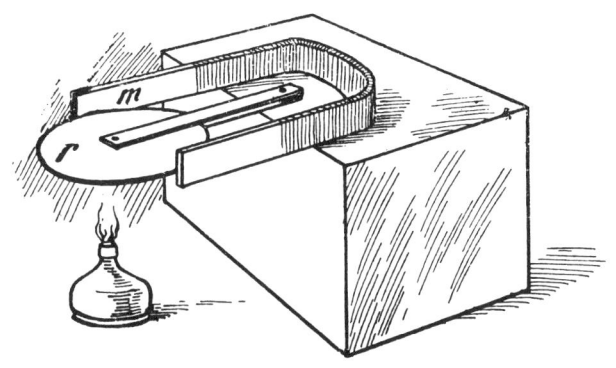

FIG. 25.

magnetised by the poles than parts adjacent to it,

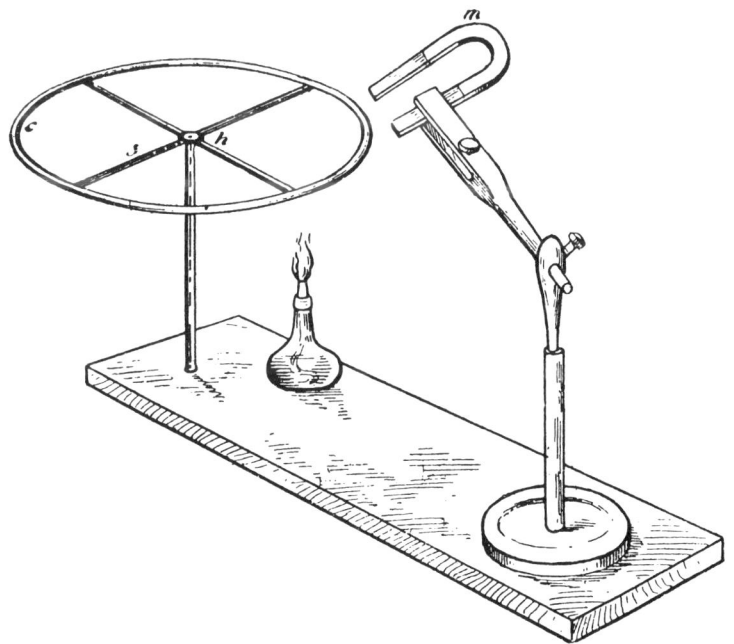

FIG. 26.

the attraction of the poles exerted on the latter will

3

be sufficient to cause a movement. If a constant heat be applied a slow rotation will result. To ensure success the disc must be sufficiently thin to prevent its acquiring a uniform temperature. Instead of a disc an iron wire bent in the form of a circle (fig. 26, c) fixed by means of four spokes, s, to a hub, h, may be used. If the poles of the magnet are too close to each other to allow of the insertion of the disc or ring, the magnet m may be so fixed as to have both poles close to and pointing towards the periphery of the disc, the heat from a Bunsen burner or other source being applied at a little distance from one of the poles.

EXPERIMENT NO. 20.

Chladni's Sound Figures on a Soap Bubble Film (fig. 27).—This is a very simple, most beautiful and instructive experiment. Cut a hole near the bottom of a small tin can, t, and fix into it a foot or more of metal or indiarubber tubing, i ; a short piece of garden hose will answer. A mouthpiece at the end of the tube is an advantage but not a necessity. Now cut a few discs, d d, of sheet metal out of a biscuit box ; these must be larger than the opening of the tin can, and be firmly attachable to the top like a cover. This can easily be done by a few radial cuts, r r, being made into the discs, bending down the strips thus obtained. The centres, c, of the discs are cut out in various shapes, such as circles, squares, pentagons, ellipses, etc. To per-

form the experiments dip a disc into a soap bubble
solution. If a film has been obtained thereby, drain
off superfluous water until the film shows the irides-
cent colours of the spectrum. Now fix the disc
firmly to the top of the tin can and bring the film
under the influence of sound vibrations, by singing
or whistling into the mouthpiece, or holding a

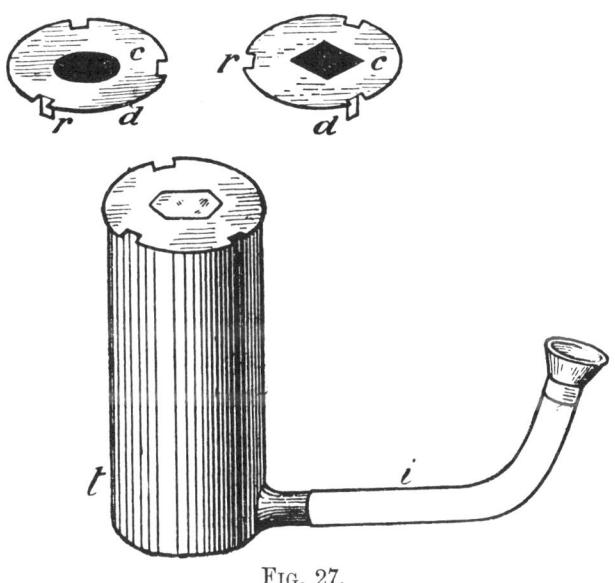

FIG. 27.

vibrating tuning fork close to it, or placing the
mouthpiece firmly against the body of a violin
while drawing the bow across a string, etc, when the
films will be thrown into vibration, producing figures
of infinite variety, and they will be found very
beautiful, though their production is obtained by
such simple means. The tin box may be fixed to a
baseboard to give the apparatus more stability, and

it will then be called Phoneidoscope. A beam of
sunlight or a parallel beam from an optical lantern
may be thrown on the film and the reflected beam
passed through a lens of 6 or 8 in. focus and received upon a white screen. Gorgeous colour effects
will be produced on the screen. To obtain a proper
image the lantern must be slightly tilted forward;
or if such an arrangement is not practicable, it may
be placed parallel with the screen, and the film
held up in the beam of light, making a slight angle
with the screen.

Experiment No. 21.

Electro-Capillary Light (fig. 28).—Send the
discharge of an induction coil through a narrow

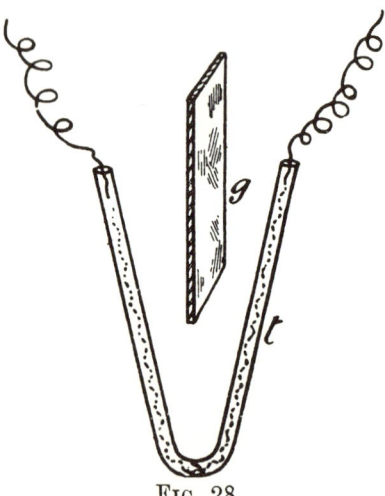

FIG. 28.

capillary tube, t,—a
thermometer tube will
answer very well—
provided with copper
electrodes and filled
with air under ordinary
pressure. The length
of the tube will depend
upon the size of the
coil; it should be a
little shorter than the
length of the spark in
the air. The discharge

through the tube will be intensely luminous, far
more so than the arc, and would be a superior

source of light if it could be made continuous.
The tube, after some time, becomes rough on the
inside and cracks; it is then useless. Variously
shaped tubes may be employed; the spark will
follow their sinuosities, but care must be taken
to leave sufficient space between the ends, as else
the discharge takes place through the air. To prevent this, however, a piece of window glass, g, may
be inserted by way of a screen between the ends of
the tube. Even if the tube be V-shaped, the discharge will follow the path of the bore.

EXPERIMENT No. 22.

A Cheap X-Ray "Tube" (fig. 29).—Cover the
top of an ordinary 16 c.p. 50-
volt incandescent lamp, l, with
a piece of aluminium foil, f,
and connect the foil with one
terminal of the secondary of
an induction coil, the other
terminal being connected with
the platinum wire of the lamp.
By means of a "screen" or
fluoroscope the bones of the
hands, feet, etc., can be distinctly seen when the latter
are placed between the screen

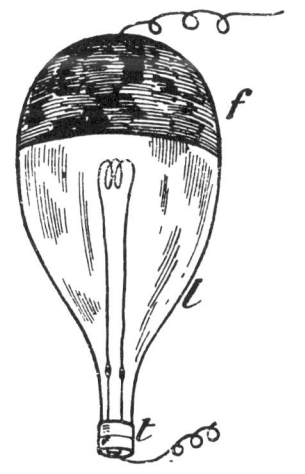

FIG. 29.

and the lamp. The lamp will, however, produce
good results for only ten or fifteen minutes, but

will recover if lighted in the ordinary way for half an hour.

Candle Burning in Water (fig. 30).—Attach a piece of metal or weight, w, of any kind, such as a piece of tinfoil, to a piece of candle that has been in use before, so that it may float vertically in a vessel of water, the latter being flush with the edge

FIG. 30.

of the candle without wetting the wick. Now light the wick; the candle will be consumed and shortened, but on account of its diminution in weight it rises in the water at the same rate at which it is consumed. This experiment affords an illustration of the law of Archimedes.

Experiment No. 24.

Electro-Magnetic Gun.—A curious instance of the tendency to complete a short magnetic circuit is afforded by the following experiment. Wind six or seven layers of insulated wire (16 to 20 gauge) round an iron tube about $1\frac{1}{2}$ to 2 ins. in length by $\frac{1}{8}$ to $\frac{1}{4}$ in. bore. Place a round bar of iron, about 1 in. in length and fitting loosely, into the tube, and turn on a strong current through the coil of the electro-magnet. The magnetic field will try to repeat itself by lengthening the iron; in fact, the piece of iron will be jerked out of the core with considerable force. Instead of an iron tube, a tube made of a piece cut from a tin can may be used. This should be rolled two or three times round a lead pencil so as to form a tube consisting of several layers of sheet iron. A current from 3 to 4 bichromate or accumulator cells is required to perform this experiment successfully.

Experiment No. 25.

Rotating Disc (fig. 31).—Wind about 30 yards of insulated wire, No 28, on a small frame, *f*, for which the outer case of a match-box may be utilised. Fix the pointed end of a needle on a small piece of wood, *w*, so that it can easily be placed inside the box. Cut a disc, *d*, of sheet iron from a tin can, a little smaller than the diameter of the box; punch

the centre of the disc so as to produce a small
indentation on which it may be suspended over the
needle point, and which allows it to rotate freely.
Now place the coil with the disc inside between
the poles of a magnet, *m,* connect the ends of the
wire with the *pillars of the contact breaker* of an
induction coil (not with the secondary) and actuate

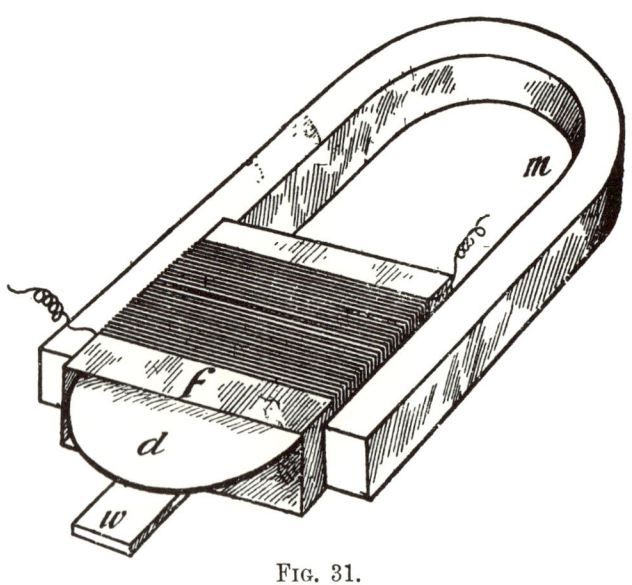

Fig. 31.

the coil with one or two bichromate cells.   The
disc will revolve rapidly, because on account of the
presence of the magnet part of the disc acts as a
magnetic needle and is deflected by the current;
but a fresh part of the iron being brought under
the influence of the magnet, this fresh portion is
deflected, and so on.

EXPERIMENT No. 26.

Fiery or Flaming Vortex Ring (fig. 32).—Cut
two lengths of thin wire of about one foot each and
bend one end of each in the form of a ring, *r*, of
about 2 ins. diameter, the other ends being connected
with the terminals, *t*, of an induction coil in such a

FIG. 32.

manner that the planes of the rings are facing each
other and about 2–3 ins. apart. Blow a soap
bubble by means of an indiarubber tubing connected with the gas supply, from a small thistle
funnel (or clay pipe), holding the latter between the
rings and slightly above them. As the bubble
enlarges it will adhere to both rings and be held by

them firmly ; the funnel can then be removed from the bubble, slightly turning it when doing so, which will facilitate the operation. The coil should now be set to work, when it will be noticed that the bubble gives off a thin cloud of steam. After a few seconds it bursts, and the sparks now passing between the wire rings, ignite the gas which passes upwards in the shape of dark-red flaming vortex ring. If this experiment is to be repeated, the wire rings must be carefully moistened and rubbed with the soap solution, or the bubble will burst on coming in contact with them. This experiment requires a coil of at least 2 ins. spark gap.

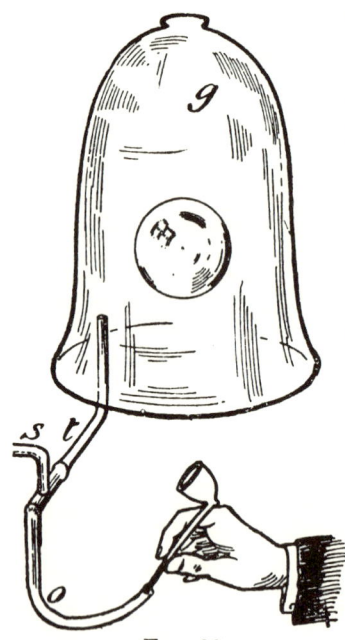

FIG. 33.

EXPERIMENT No. 27.

A Floating Soap Bubble (fig. 33).—This experiment, devised by the author, is a modification of the well known experiment of floating a soap bubble in a basin containing carbonic acid gas or vapour of sulphuric ether. Invert and fix a large glass globe, *g*, a foot or two above the table and fill it with ordinary coal gas. This can easily be done by means of indiarubber tubing, *t*, connected with a gas burner, fixing the free end of the tube against the inside of the inverted globe *g*.   Now blow a soap-bubble at the end of a clay pipe or glass tube likewise connected with the gas supply, *s*, by means of india-rubber tubing, *o*.   When set free this bubble will rise in air, and if released under the inverted glass globe it will rise inside the globe, but only to that part where it encounters the gas, and will remain there stationary.   Hydrogen may of course be used instead of coal gas.

EXPERIMENT NO. 28.

Modification of Foucault's Pendulum Experiment to prove that the earth turns round (fig. 34).— Fill a large bowl with water and put it on the floor of a quiet room not subject to being shaken.   Sprinkle over the surface a thin layer of either lycopodium or sifted resin.   Next trace a line in a similar manner straight on this layer by means of powdered and sifted

FIG. 34.

coal.   Make a mark on the edge of the basin opposite the line.   In a few hours the line will apparently have shifted its position though not in reality, for the basin has partly turned with the building, leaving the water stationary.   It will be found that the black

line turns from east to west, which proves that the
basin and building have turned from west to east.

EXPERIMENT No. 29.

Pea suspended in Air (fig. 35).—Stick two pins
on opposite sides into a hard pea so that their
points are directed to the centre, and bend them
slightly downwards. Now put the pea on the open
end of an indiarubber or glass tube, *t*, with a bore
slightly less than the diameter of the pea. Blow

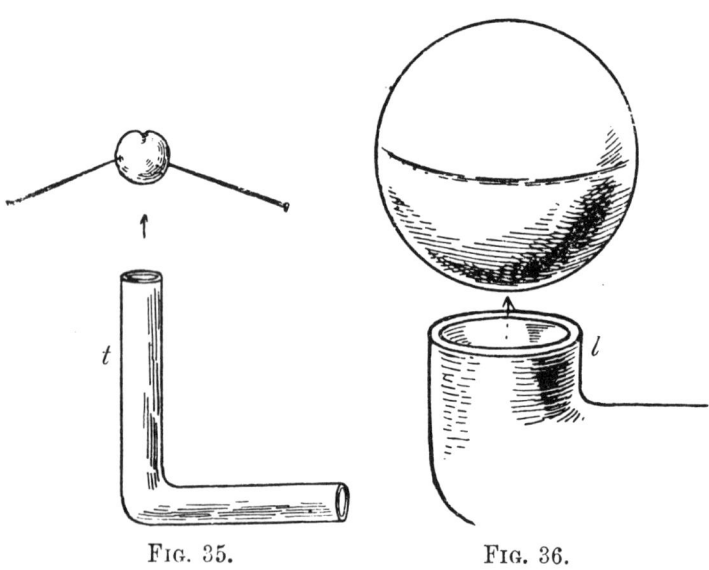

FIG. 35.　　　　　FIG. 36.

strongly but evenly into the other end of the tube,
when the pea will be seen to rise two or three inches
and remain stationary above the tube. If the force
of the blast be gradually diminished, the pea will
resume its place on the tube. This experiment

may also be performed with a small celluloid ball
(fig. 36) (obtainable in any toy shop), but a larger
tube, l, with a bore of about $\frac{1}{4}$ of an inch is required.
No attachment to the ball is needed. If an arti-
ficial and continuous blast is applied, such as obtainable from a wind chest or foot blower, the ball will
rotate at a certain unvarying height above the
opening of the tube, the height depending upon
the force of the blast.

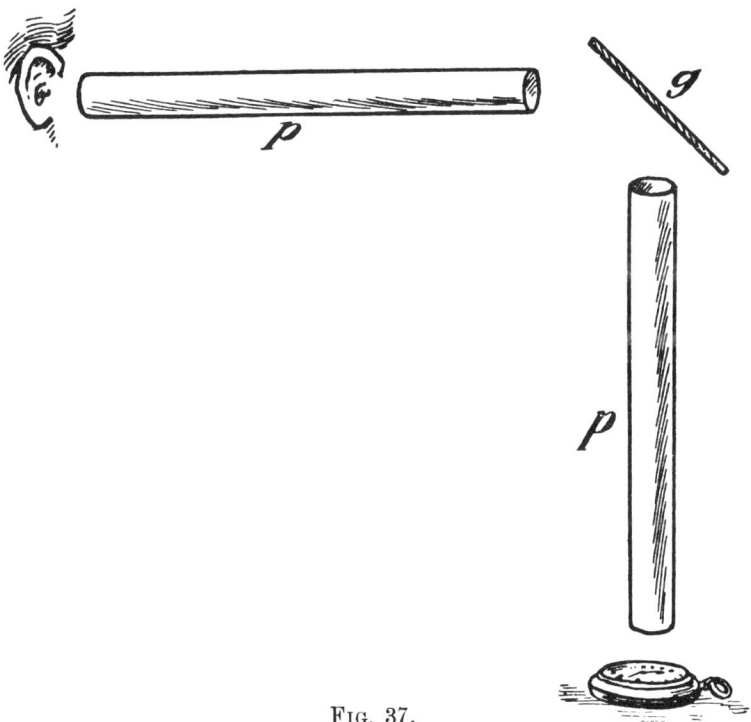

FIG. 37.

EXPERIMENT No. 30.

Reflection of Sound (fig. 37).—Make two paper

tubes, each 3 ft. long by 3 or 4 ins. in diameter, $p\,p$; place them at right angles to each other on the table so as to form two sides of triangle. Now, placing a watch at one opening no sound will be perceived at the other opening, but if a piece of glass or polished board, g, be placed at the meeting-place of the two tubes and in such a position as to form an angle of 45 degs. with each, the ticking of the watch will be distinctly audible.

EXPERIMENT No. 31.

High Frequency Currents (fig. 38).—These can be easily produced by means of an induction coil, two Leyden jars, and two universal dischargers. The jars, $j\,j$, are placed in front of the coil, c, on a piece of glass, g, to insulate them, and the inner coating of each is connected with one of the brass balls, $b\,b$, of a discharger which in its turn is connected with one of the poles, $p\,p$, of the coil. The outer coating of each jar is connected with one of the balls, $d\,d$, of the second discharger. Between the two jars is placed a coil of very thick copper wire, w, of about eight or ten turns, the ends of the coil in contact with the covering of the jars. If, now, the coil is thrown into action, sparks will pass between the balls of both dischargers. Though in appearance very much alike—thick and noisy— they differ considerably in character : for whereas the discharge between the inside of the jars if received by the human body would have a disastrous effect, the discharge between the outer covering of the jars produces no such effect at all ; in fact, the balls of the second discharger may be freely and fearlessly grasped though a torrent of sparks is passing. The

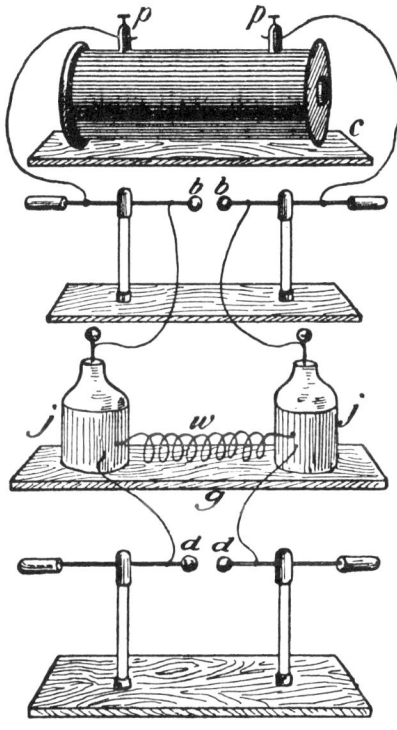

FIG. 38.

absolute freedom from all sensation is at first quite surprising, especially to lookers-on ; and it is difficult to persuade people to try the experiment and judge for themselves of the absence of any perceptible physiological effect of these high frequency currents.

EXPERIMENT NO. 32.

An " Accumulator " for Static Electricity (fig. 39).—Make an ordinary Leyden jar, *j*, with metal rod just long enough to reach about as far as the middle of the jar. Roll up a piece of stout tin-foil or thin sheet iron (cut from a canister), so as to form a tube, *t*, which will slide easily to and fro on the rod inside the jar. A small ball, *b*, must be attached to the end of the tube, which should be long enough to slide down and touch the bottom of

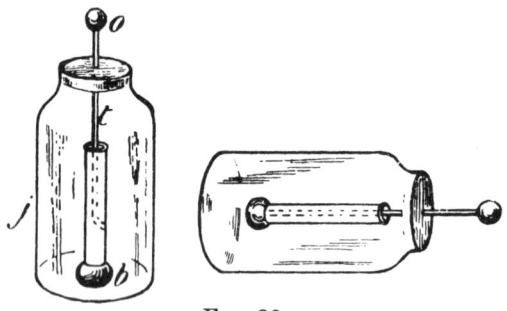

FIG. 39.

the jar when the latter is standing upright; but on being turned upside down it should slide back and break the connection between the rod and the inner covering. Place a small quantity of chloride of lime in a small box with a perforated lid and cement the box to any part of the inner covering of the jar. The lid of the jar must now be hermetically sealed to the mouth of the bottle, and the hole through which the rod passes must likewise be made quite air-tight. Charge the jar in the ordinary way, then turn it upside down so as to allow the tube to slide up to the lid, and lay the jar on its side on some insulating material such as glass or vulcanite. The jar will retain its charge for many days, and be found useful for obtaining small quantities of electricity from the outer ball, o, by simply allowing the tube to come in contact with the inner covering and then to slide back against the cover of the jar.

FIG. 40.

EXPERIMENT NO. 33.

Decomposition of Steam by the Secondary Current from an Induction Coil (fig. 40).—The cork, c, of an ordinary flask, f, for boiling water, has,

4

besides the delivery tube t, two wires, w, inserted in it which project a short distance into the neck of the bottle. The outer ends are connected with the poles of the secondary of a coil. The delivery tube must be bent in such a way as to allow the gas evolved to be collected under water in a gas cylinder, c. When the water in the flask is boiling the current is turned on. The steam is decomposed, and both hydrogen and oxygen are given off at both wires, and escape, together with some steam. The latter, however, is condensed again on coming in contact with the water of the collecting apparatus, and the hydrogen and oxygen rise in small bubbles from the end of the delivery tube.

EXPERIMENT No. 34.

Experiment with Toy Balloons (fig. 41).—It is well known that an ordinary piece of brown paper

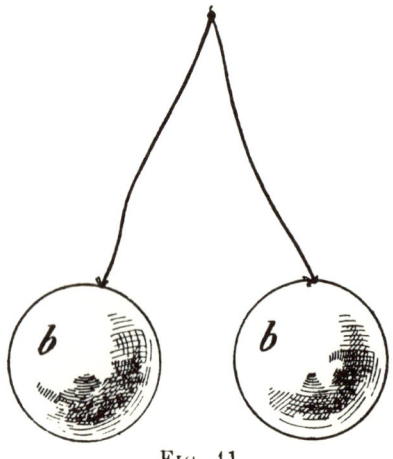

after having been dried before a fire and rubbed with the dry hand, a cloth brush or a silk handkerchief, will strongly adhere to a wall and remain in contact with it for a long time. An ordinary toy balloon filled with air will in a similar manner cling to the

FIG. 41.

wall or ceiling after having been stroked with a piece of fur. Two balloons, $b\,b$, suspended from silk threads and electrified by rubbing with fur or wool, will repel each other after the manner of pith balls.

Experiment No. 35.

Chladni's Figures on Membranes (fig. 42).—Chladni's figures are generally produced on discs of glass and metal fixed horizontally in a vice and sounded by means of a violin bow. They can, however, be quite readily produced on a membrane fixed drumlike fashion over the end of a metal or wooden

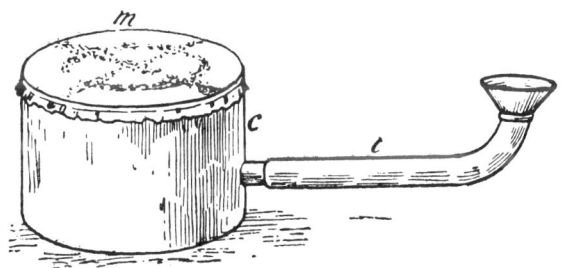

Fig. 42.

cylinder. An ordinary small tin canister, c, with a piece of sheet of indiarubber, m, stretched over the opening, will answer very well. The closed end of the canister is connected with a short length of indiarubber tubing, t. To produce the figured lines, sand or lycopodium is thinly scattered over the membrane. The powder will arrange itself in symmetrical figures according to the pitch of the note sung into the tube. To produce the best effect the sound of the voice should subside gradually, without, however, changing the pitch. This needs a little practice. A small projecting edge of paper pasted round the circumference of the membrane will be found an advantage, as it prevents the powder from being jerked off.

EXPERIMENT NO. 36.

Magnetic Dipping Needle (figs. 43–47).—A good dipping needle is an expensive piece of apparatus, but a very effectual needle may be constructed of

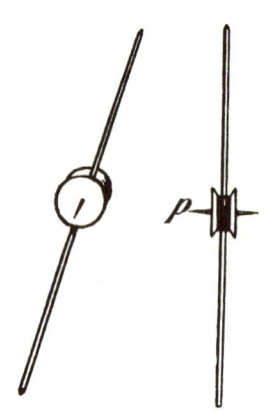

a knitting needle and two drawing pins. A stout knitting needle should be used for that purpose, and care should be taken to select one perfectly straight and non-magnetic. Roll the needle over a perfectly smooth surface, such as a piece of plate glass; if the needle is at all bent the unevenness will become apparent at once. To find out whether the needle is already a magnet, each

FIG. 43.

end of it should be approached alternately to the N. and S. poles of an ordinary compass needle. If both poles of the needle are attracted by both ends of the knitting needle, the latter will be suitable for our purpose; but if any repulsion takes place, the needle should be rejected. Fix the drawing pins by means of a pellet of beeswax to the middle

of the needle on opposite sides, points outward, as
shown in fig. 43, *p*. Bend a piece of thin metal
(not iron) plate in the form of a horseshoe (fig. 44)
and suspend the needle from it as shown in the
figure. If one half of the knitting needle is found
to be heavier than the other, the needle will of
course assume an upright position, and in that case
it should be pushed through the wax until both
halves are exactly of the same weight. With a

little patience the needle can
soon be so adjusted as to remain
in any position between its points
of suspension. Care must of
course be taken to place the
drawing pins in exact juxta-
position to each other, their
points forming a straight line
which passes through the middle
of the knitting needle. If the
needle has been satisfactorily
adjusted it may now be mag-

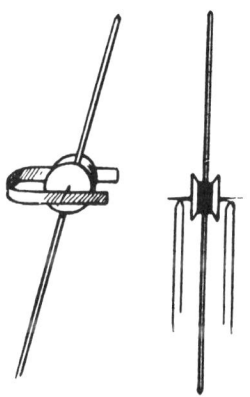

FIG. 44.

netised by any of the methods described in text-
books. The author prefers placing it in a coil of
wire and passing an interrupted current of three
or four bichromate or accumulator cells through it
for a few seconds. This method prevents the for-
mation of consecutive points, and ensures the needle
being magnetised in the direction of its own axis
and not obliquely, as might be the case if single
or double touch by magnets is used. The needle
having been properly magnetised may now be

poised again—N. pole to earth—on the metal fork
as shown in fig. 44.   This suspension is very
delicate, and the needle will now be found very
sensitive.   It should oscillate freely, and ultimately
come to rest at an angle 67 degs. if placed due
north and south.   If placed in the direction of
east and west the needle should assume a vertical
position.   The reasons for these positions are given
in text-books.   Instead of poising the needle on the

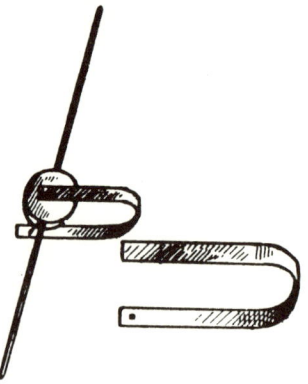

FIG. 45.

fork, a slight indentation may be made inside each
prong for the points of the drawing pins to fit in
(fig. 45), but care must be taken not to allow more
than a minimum of pressure on the points to keep
the needle in place, lest the friction between the
fork and the point should impede the free movement.   A scale may be gummed to one branch
of the fork, or a small mirror may be fixed to the
S. pole of the knitting needle before adjusting and
magnetising it, in order to throw a speck of light

on to a scale on a screen. These minor details require no description. In order to mount the needle, the bend of the fork may be cemented to a glass tube (fig. 46, *g*) which is bent at right angles and the free arm of which is inserted in the open end of an upright glass tube, *t*, of a diameter just large enough for the first tube to pass

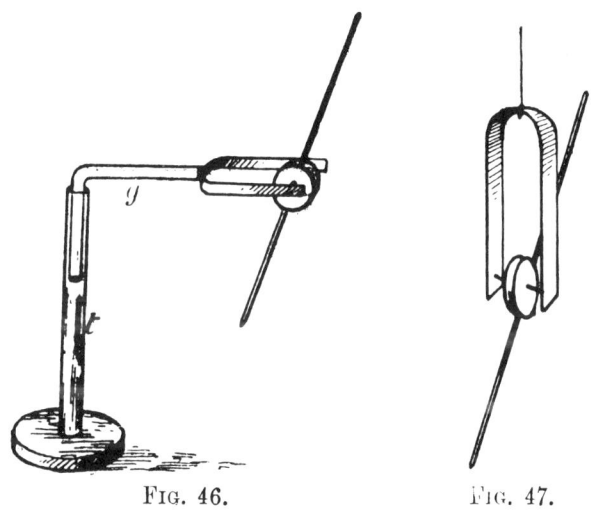

FIG. 46. FIG. 47.

in (fig. 46). This arrangement will allow of a free and easy rotation of the needle, showing its dip in all the points of the compass. If the needle is suspended from a thread fastened to the bend of the fork it will assume an N. and S. position of its own accord (fig. 47), showing both declination and inclination. It should be borne in mind that such a needle is much more sensitive than an ordinary compass needle, and it should be placed at least three feet away from any iron or steel.

EXPERIMENT No. 37.

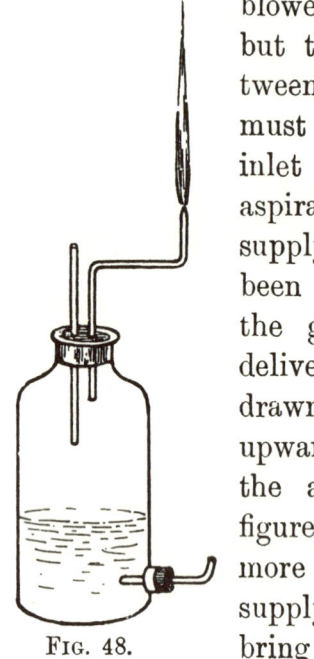

FIG. 48.

Sensitive Flame (fig. 48).—A good sensitive flame can be produced by means of the aspirator and blower described in Experiment 17, but the perpendicular distance between the two pieces of apparatus must be not less than 8 ft. The inlet tube, r (fig. 22), of the aspirator is connected with the gas supply, and when the water has been allowed to flow for some time the gas may be lighted at the delivery tube $b\,t$, which must be drawn out to a fine point directed upwards. The flame should have the appearance as shown in the figure, 48, and be three to four or more inches in length. The water supply must be regulated so as to bring the flame near the point of roaring. It will then be sensitive and bob up and down at the least noise.

EXPERIMENT No. 38.

Curious Phenomenon caused by Moist Aluminium in Contact with Mercury (fig. 49).—This phenomenon was accidentally discovered by the author in the year 1895, and has often been shown by him to people interested in scientific matters and at public lectures. The aluminium used for the experiment was always obtained by *breaking off* a small piece from an aluminium penholder, so as to obtain a very clean and rugged edge, without which the experiment will not succeed. Thoroughly moisten the piece and then dip it in a small quantity of mercury, moving it about for a few seconds. In about a minute or two small white points will appear at the sharpest corners, and these points will rapidly and visibly increase, appearing in every way like a vegetable growth. The movement and gradual increase of this feather-like growth can be distinctly seen, and forms a most interesting object when projected on a screen. The author has frequently obtained thick feathery growths of quite an inch in length. If a few sparks from an induction coil

FIG. 49.

are allowed to pass between the aluminium and mercury, when these two metals are brought in contact the feathery growth will take place much more rapidly and profusely. Care should be taken to thoroughly moisten the aluminium.

EXPERIMENT NO. 39.

Experiment with Celluloid Ball (fig. 50).—Float an ordinary celluloid ball, *b*—such as is used for

table-tennis—in a basin of water, and direct on to it
a gentle stream of water, *j*, from an india-rubber
tube, *t*. The ball will cling closely to the jet of
water and follow every movement of the jet, though

FIG. 50.

the latter may be moved about very freely. This
experiment is an excellent illustration of surface
tension. If the jet of water is allowed to escape
under too much pressure the experiment fails.

EXPERIMENT NO. 40.

Ball and Nozzle Experiment (fig. 51).—Make
a conical nozzle, *n*, of a piece of sheet-metal large
enough to allow a light ball, *b*—a celluloid ball such

as can be bought in toy shops will do—to fit into
it. Connect the nozzle with the water supply and
bring the ball in front of the issuing jet of water.
The ball will be sucked into the nozzle and held

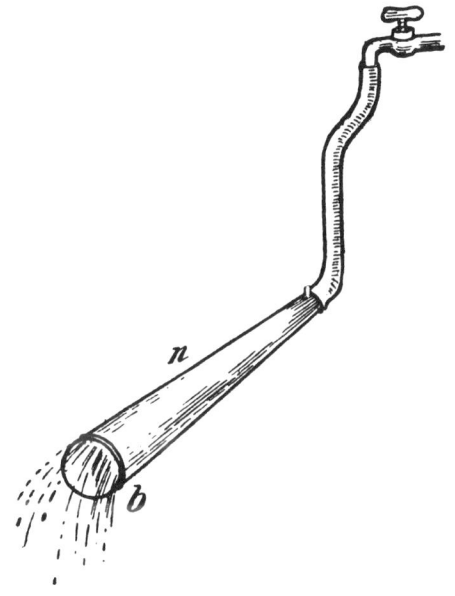

FIG. 51.

there by the pressure of the air, but the water will
continue flowing all round the ball. The nozzle
may be held in any position with the same result.
A strong blast of air will have the same effect.

EXPERIMENT No. 41.

A Top Propelled by Air (fig. 52)—This is an
experiment of a similar nature. Cut a cardboard
disc, *d*, about 4 ins. in diameter. Mark out in

pencil a series of parallelograms, *p*, obliquely to the radii of the disc, as shown in the figure. With a sharp penknife cut through one of the longer, *l*, and two of the shorter sides, *s s*, of each parallelogram, and turn up the cardboard thus detached at right angles to the surface of the card. From the other side force a common pin through the centre of the disc, leaving a space of about ¼ of an inch between the head of the pin and the card. This pin forms the pivot of the top. Now insert the pointed end of the pin loosely into the bore of an ordinary

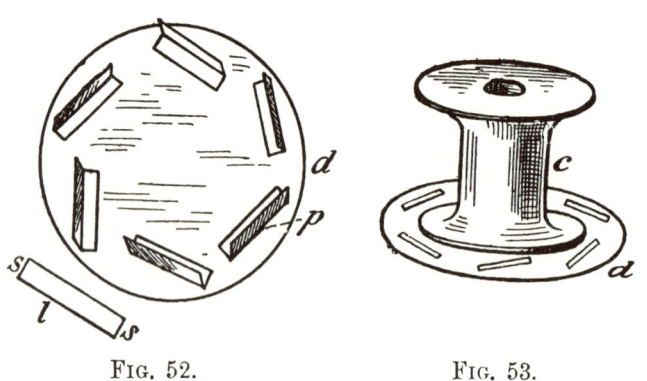

FIG. 52.   FIG. 53.

cotton spool, fig. 53, *c*, hold the disc up lightly against the spool, and blow into the open end of the bore. The top will spin rapidly, and although no longer held up by the hand it will retain its position, as the stream of air issuing between the disc and the spool creates a partial vacuum and the atmospheric pressure on the under-side sustains the disc. If the blowing ceases, the top drops, but continues to revolve on its pivot.

EXPERIMENT NO. 42.

Thermo-Electric Currents (fig. 54).—Fix a slip of sheet-copper, *s*, horizontally, and heat it to a red heat by means of a Bunsen burner, *b*, placed underneath. Firmly press the end of a thick copper

FIG. 54.

wire, *w*, against the upper surface of the copper plate. Both the plate and the wire are connected with a galvanometer or voltmeter. The current obtained has an electromotive force of from $\frac{1}{10}$ to $\frac{1}{5}$ volt.

EXPERIMENT NO. 43.

Experiment with the Induction Coil.—Mix a small quantity of finely divided metal, copper or zinc, such as used by color printers or for making gold and silver paint, with a little lubricating or colza oil, and spread the mixture between two pieces of glass, inserting at the opposite sides slips of tinfoil to be connected with the poles of the secondary. There should be sufficient metal dust in the mixture to make the layer of oil almost opaque. On turning on the current a sudden movement will be perceptible in the mixture, and it will become quite transparent, resuming its former opacity when the current is stopped. It must be assumed that the metal particles when at rest are lying on their flat sides, and that under the influence of the current they assume an upright position, allowing more light to pass through the spaces between them. If the slips of tinfoil are brought sufficiently close together to allow sparks to pass, this discharge will take place silently and the presence of ozone will be noticed. In the dark the forked discharge affords a beautiful spectacle.

EXPERIMENT No. 44.

Interference of Waves (figs. 55 and 56).—This experiment is generally performed by means of two tuning forks, but the arrangement described below will be found to possess many advantages, and can be performed with sun or artificial light, and only requires a corner of the room shaded from strong light for projecting the image. Two flat steel springs, 5 or 6 inches long, are arranged as shown in the figure. The spring a is passed through a slit in the board, t, so as to allow it to be raised or lowered at pleasure. The other spring is bent at

right angles, the shorter arm being screwed to the
table and making an angle of about 45 degrees with

FIG. 55.

the opposite spring.   To the top of each spring a
small mirror, cut from a looking-glass, is firmly

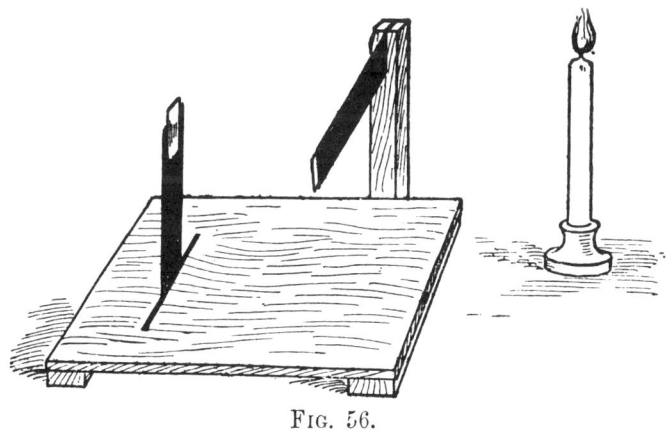

FIG. 56.

cemented, the one attached to spring a being larger
and projecting as shown in the figure, so as to keep

the mirror of spring C—even when both are vibrating—within the limits of a beam of light reflected from the large mirror. Light from a source placed opposite spring a will be reflected on to the mirror of spring C, and thence to a dark wall. The proper position of the source of light must be found by turning the table until a double reflection is obtained and a spot of light appears on the wall. If, now, one spring only is vibrated a straight line is projected; but if both are vibrated simultaneously, a figure, the combination of two vibratory motions, will be produced. By altering the length of spring a, various effects will result. The little table should be placed on some books or raised otherwise, so as to allow the spring to be lowered to any extent. If the springs should be preferred at right angles to each other, the spring C should be fixed to an upright post provided with a slit for the spring to be slipped in, as shown in the figure 56. A long-focus lens may be placed between the spring and the wall, so as to bring the beam of reflected light to a focus and increase the sharpness of the outline of the figures

EXPERIMENT No. 45.

Curious Experiment in Capillary Attraction (fig. 57).—Fill two wine-glasses (or tumblers), g, of the same size with water by immersing them in a basin of water, join their rims accurately, and remove them from the basin. If the rims fit accurately upon each other, the water will remain in the

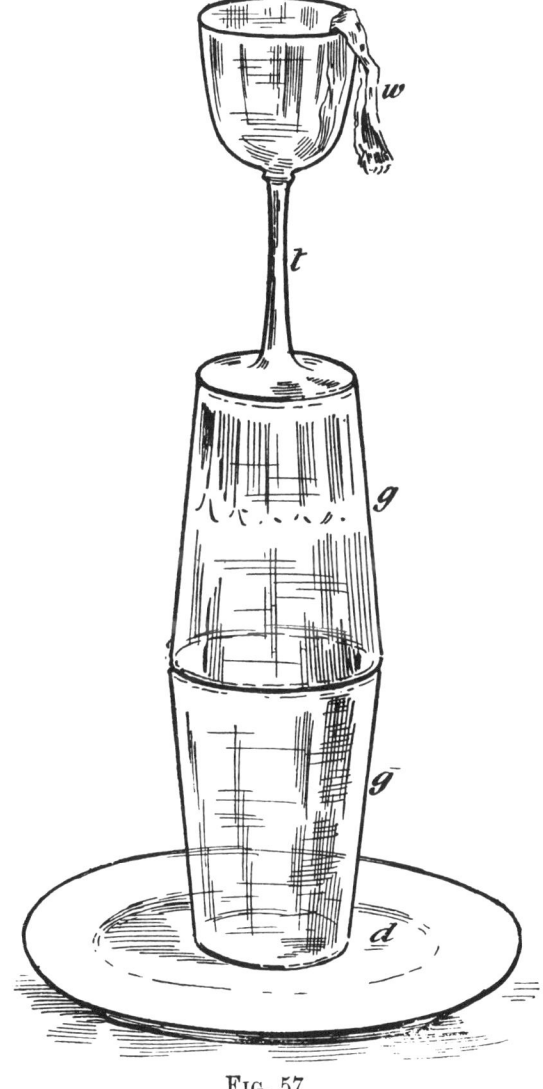

glasses.   They may now be placed upright on a dish,

d, and a third glass, *t*, put on the top. Fill this glass with port or strong claret, or, better still, with spirits of wine in which a little aniline has been dissolved. Over the edge of the top glass put a strip of wool or cotton, *w*, previously moistened in the liquid. This will act as a syphon, and allow the contents of the top glass to flow away in drops, falling and running down on the outside of the glass underneath until they arrive at the junction of the rims, when they are drawn in by capillary action, and being specifically lighter than water, rise to the top of the upturned glass. Of course, the displaced water finds a way out between the rims, and flows to the bottom. This action goes on until the top glass is emptied, and the whole contents of it have found their way into the inverted glass; the water in the lower glass remains perfectly clear.

EXPERIMENT NO. 46.

An Artificial Sky (figs. 58 and 59).—A modification of the late Prof. Tyndall's experiment to show that the blue color of the sky is due to exceedingly minute particles of matter suspended in the air. Make a box of six pieces of window glass, *b*, of 8 or 10 ins. square each, by gluing narrow strips of black paper along the edges, the black side turned towards the box. Before fitting in the last piece a small quantity of vaseline or glycerine must be equally spread over the inside of one of the squares. Two of the squares opposite to each other should each have one of their corners broken off to allow of the introduction of two pieces of ordinary quill glass tubing, *q*. These should be fitted air-tight by means of sealing-wax or cycle cement. To prevent access of air into the box the paper joints must be varnished once or twice. A foot of india-rubber tubing may now be slipped over each of the glass tubes and closed by means of pinch-cock *p*. The box must now be set aside for at least twenty-four hours, for all the dust in it to settle down on the

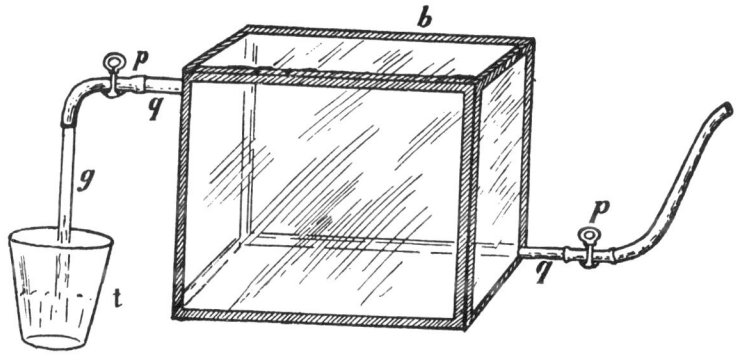

FIG. 58.

vaseline at the bottom of the box. A beam of light, *l*, from an optical lantern, *o*, passing through the box will be visible before entering and after leaving the box by the light being reflected from the numerous motes floating about in the air; inside the box, however, it will be quite invisible on account of the absence of floating particles in the enclosed air. One of the india-rubber tubes is now connected with a convenient length of glass tubing, *g*, containing a few inches of cotton wool firmly

rammed together. To prevent any dust that might
have settled down in any of the tubing from finding
its way into the box on opening the pinch-cocks,
it is advisable to pour some glycerine through them
before placing them in position, otherwise the experiment is likely to be a failure. To perform the
experiment we want to introduce a very small
quantity of minute particles of matter into the box,
and this object is best attained by lowering the

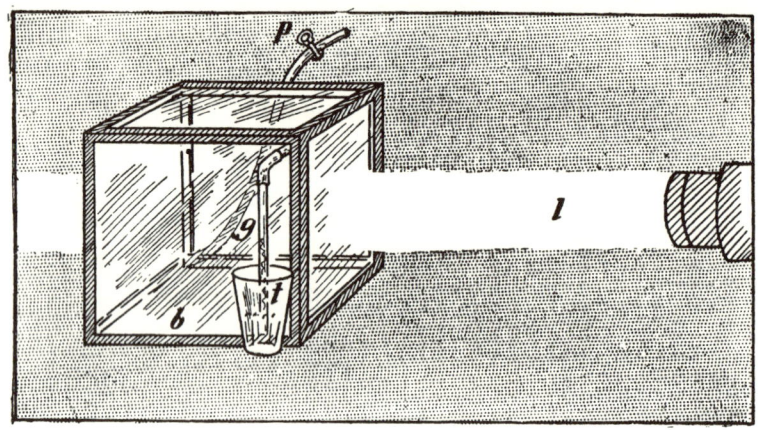

FIG. 59.

glass tube with the cotton into a beaker or tumbler, *t*,
and blowing a puff of tobacco smoke into the latter,
and then opening the pinch-cock above. We shall now
want to allow a trace of the smoke to enter the box ;
this can be achieved by taking the india-rubber tubing
on the opposite side into the mouth, sucking the air,
and simultaneously opening the second pinch-cock.
Great care must be taken not to draw in too much
smoke and to shut the pinch-cock again at once.

The smallest quantity that can be perceived entering the box will be sufficient, and as it is slowly diffused in the box a beam of strong light will show it to be of a bluish tint, the colour of the sky. If the light appears greyish or white it is a sign of too much smoke having been admitted. In Tyndall's original experiment air is drawn into a large glass cylinder through a piece of very fine platinum wire gauze which is kept at a white heat. Here the larger dust particles are all burnt and only the very smallest escape and pass into the cylinder. The modified experiment described above is more readily performed, nearly as effective, and requires only a small outlay. If not successful the first time, the experiment should be repeated until a satisfactory result is obtained. With care and skill it is certain to succeed. The author has repeatedly performed it at lectures.